Instagram

Master Instagram Marketing – Instagram Advertising, Small Business and Branding

Grant Kennedy

Contents

Page

Introduction...9

Chapter 1 – Overview of Instagram............................11

Chapter 2 – Origin and Basics...................................15

What is Instagram and how did it Begin?...............15

Instagram Terminology...16

Setting Up Your Instagram Account for Business..18

Making Your Account Look Attractive to Get More Followers...22

How to post on Instagram.......................................29

How to enrich and nurture your community 31

Chapter 3– Why Instagram is important for your business?...35

Chapter 4 – Using Instagram for Business................37

    Determine Your Objectives......................38

    Develop a Content Strategy.....................40

    Making content discoverable on Instagram..........43

    Build Content Themes..........................44

    Determine Types of Content.....................45

    Set a Content Calendar, But Be Flexible.............45

    Consider Curating User-Generated Content..........46

Chapter 5 – Making Your Posts Profitable................49

    Gaining More Visibility Via Facebook....................51

    Gaining More Visibility Via Your Official Website..51

    Gaining More Visibility Via Hashtagging...............52

    How to use Instagram Direct......................52

Chapter 6– Hosting an Instagram Contest or Giveaway...............................57

    Contest #1 – Random Selection Contest...............60

    Contest #2 – Maximum Likes Contest..................60

    Contest #3 – Walk In to the Store Contest.............61

    Contest #4 – Website Contest.....................62

    Contest #5 – Video Contest......................63

Contest #6 – Shout-out Contest............................64

Chapter 7– Establish Clear Guidelines.....................71

Create an Instagram Style Guide...........................71

Chapter 8- Foster Engagement & Set Guidelines for Community Management...77

Optimize your bio & link....................................77

Following Accounts..78

Monitoring Hashtags..78

Search & Explore...79

Analyze Your Results..79

Chapter 9- Apps to help you boost you Instagram Marketing..87

Iconosquare..88

Hyperlapse...88

Offerpop..89

Slidagram..89

VSCO Cam..89

Afterlight..90

Followgram..90

Chapter 10- The 5/3/1 Rule....................................91

Instagram Etiquette..92

Tips & Tricks..94

Instagram + Beyond................................95

Leveraging the Instagram......................96

Using the Instagram brand and trademark............97

Using Instagram screenshots.................97

Chapter 11- Some Quick Marketing Tips for Instagrammers..99

Chapter 12- 20 Mistakes you can't afford to make in Instagram Marketing................................105

20 Mistakes...105

Conclusion...111

Introduction

The Twenty First Century is the Age of the Digital Media. In the fast paced, high powered world of today, if you do not exist online, you do not have any presence in the world at all! Social media has become not only a platform to meet new people and connect with them, it is vital if a person wants to be able to have an active, open life.

When it comes to running a business, these social media networks become even more important! And for a small business, it could prove the difference between making a profit and running into a loss! The internet is a worldwide tool; making full use of these networks allows you to gain access to customers on the other side of the world that you normally would not even be able to speak to! So keeping up with the latest trends on social media is an essential tool of marketing that you cannot afford to miss.

As an entrepreneur, you need to grow your fan following in order to grow your business. Social media platforms are a great help as people these days want visual content! Instagram is the best place for that. Instagram marketing will save you time and money, and it's a great way to build audience. And of course, you will get to interact with the brand as well.

For growing your followers and brand name, you need to create attractive contents as well; so that people check back on your profile – revisit again and again!

So, do you know how to use Instagram to market your business? Do you get enough interaction with your brand? Do you know how to get more followers on Instagram?

Instagram is a great way to connect with your consumers who are socially engaged. Instagram has more than 400 million active users every month and 3.5 billion photos are liked every day. Use Instagram for business purposes in the right way, and you could have an instant viral marketing success. Use it wrong, and your efforts on the site could be a big empty fail.

Well, that's what this book is all about! In the following chapters, I have elaborated on what Instagram is, how you can use your Instagram to gain more visibility for both yourself as well as your business and how you can market your business through this platform and augment your overall marketing strategy.

Thank you for choosing this book, I hope you find it useful!

Chapter 1

Overview of Instagram

Instagram is an online photo/video sharing and social networking service, currently the fastest growing social platform globally since its launch in 2010. It was acquired by Facebook in 2012 and is reportedly the social network that accumulates the most engagement.

The network is predominantly used via a mobile app available on iOS and Android devices, allowing users to take and share both photos and videos from their smartphone, follow and be followed by other users, as well as like and comment on each other's posts. A web version is also available, however at this time browsing and uploading features are not present, so it is only possible to view your own profile or profiles that you already know the URL for.

When an image or video is published, it will appear on the users profile and their followers newsfeed, as well as being given the option to share with other social networks such as

Facebook and Twitter. Alternatively, you can select specific people who you would like to receive your publication, rather than sharing it with the public or your followers.

There is the option to keep your profile as private, however on public view, hashtags, much like on Twitter and Facebook, can be added to posts for discovery in searches and for use when browsing other users' content.

Instagram's key defining features are the filters available, adding a digital layer that gives the impression of professional photography and it's trademark square shape that it confines photographs to, resembling a Polaroid or Kodak Instamatic image.

Instagram has grown enormously since its 2010 launch, reaching over 300 million active users each month, the network now worth a massive $35 billion. With over 20 billion photos having been shared to-date plus 8,500 likes and 1,000 new comments per second, it's easy to understand how active Instagram is.

As well as for personal use, Instagram has also been adopted by brands as part of their marketing strategy. When compared to other social networks such as Facebook and Twitter, Instagram engagement with brands was over four times that of the other platforms.

A 2014 survey found that 49.8% of millennial Instagram users rank Instagram in their top 10 channels to discover new products and 58.6% report being more likely to remember a great brand on Instagram over the brands on television commercials, demonstrating its power as a marketing tool.

In addition to this, since Instagram videos were introduced in June 2013, 40% of the most-shared videos were created by brands. This may be held accountable to the human brain processing visual content 60,000 times faster than written content, making Instagram the perfect tool for marketing.

Over 90% of Instagram users are under 35 and are made up of mostly women (68%). Instagram now has more than 200 million active users uploading 60 million photos/videos everyday. More importantly, 70% of them check Instagram at least once a day and 31% of them earn more than $50k a year.

Top 5 countries with the most users (Statista):

1. USA
2. Brazil
3. Russia
4. UK
5. Canada

Chapter 2

Instagram – Origin and Basics

Before we begin to learn the ins and outs of the app itself, let me give you a quick rundown of the app's origin story.

What is Instagram and how did it Begin?

Invented by Mike Krieger and Kevin Systrom, Instagram initially began as an app for iPhones called Brubn. As the app evolved over it time, its focus kept shifting until its creators decided to repurpose it for one thing – sharing mobile photography. The name Instagram itself comes from its origins – given that it is very similar to an instant telegram, the two inventors mashed up the two words together to give you Instagram.

The app was geared towards only iPhone Users at the beginning; it did not take long before an Android version was released! Given how popular the app was becoming, other big names in social media like Facebook began to sit up and take note; Facebook, in fact, approached the owners and started talks to acquire it. A couple of months later, the deal was well and truly sealed, with the app being appraised at about a billion dollars or so!

Over the years since its launch, Instagram has constantly been updated to keep its hold over tight over its users; for instance, in the tail end of 2013, Instagram added the Direct Messaging feature to its uses so that it could allow for its users to chat with each other directly, despite being a photography app first and foremost. The app keeps making changes and offers newer updates every once in a few months to keep things interesting and fun!

Instagram Terminology

As with most social media sites, there are some terms that you will need to know in order to really grasp Instagram. To make it easy for you, here's a quick list of some of the lingo you'll want to know.

You should also be aware that people who use Instagram love to put "insta" in front of everything. You'll see "instagood" or "instasuccess" and other similar terms frequently.

App – Abbreviation for "application". It is a software application downloaded to a mobile device that provides access to the Instagram site.

Feed – The collection of current posts shared by those you follow. This is designated by the house icon on the bottom navigation panel.

Filter – The editing feature which can be applied to your post to enhance features and colors.

Followers – The people that follow an Instagram user.

Following – The people that an Instagram user follows.

Frame – Certain filters for photos can be further enhanced by adding a border (a frame) to the image.

Gallery – The collection of a user's Instagram posts.

Hashtag – The # symbol placed in front of key words. Hashtags should be key words or phrases relevant to the description of the Instagram post. Hashtags are searchable on Instagram.

Home – The Home screen is the list of activity of all the users you follow. Also referred to as the feed.

Instagramers – Instagram users. A collective term used to refer to people who use Instagram.

Instameet – A gathering or meeting of local Instagramers to take Instagram photos.

Instavideo – Often used to describe the Instagram video feature.

Latergram - Something you post on Instagram at a "later" time.

Like – The measure of appreciation of a post is documented in the number of "likes" it receives. A "like" is represented by a heart on Instagram. You can like a post by double-tapping the image or tapping the heart below the caption.

Post – Any visual content uploaded to an Instagram profile. Posts can consist of photos or videos.

Profile – This is your account information on Instagram. Your profile consists of your name, username, profile photo, and your photo gallery

Username – The name a person uses on Instagram to define their profile address. This may be any configuration and does not have to relate to their actual name.

Setting Up Your Instagram Account for Business

It goes without saying that the first thing you need to do is to download the app. It can be accessed through a web browser, but it is, first and foremost, a mobile platform, so be sure to download it on your phone! Step number two is to register your account – as a business owner, it's probably best you sign up for two accounts. Your personal account is for you and your social life, but the public account is the one you will be using to promote your company. It goes without saying that your username should be something closely related to your company's name.

When setting up your Instagram profile, It's crucial that it's both easy to find and gives a positive representation of your brand.

Firstly, you need to make sure that your username reflects your company. When signing up, make sure your company's name is in your username and in the case that you want to create multiple accounts for different branches, you may also want to include your country/region in the username too. If your brand is present on multiple social media platforms, using the same username for each account will make it easier for people to find you.

In fact, unless the name has already been taken by another user, keep only your company name as your username. Otherwise, keep it as close to it as possible. For instance, if you run a high end boutique named *Awesome Girl Apparel*, try combinations using this name only, like *Awesome_Girl_Apparel* or *AwesomeGirlApparel*, etc. This way, your prospective customers can find you easily on Instagram.

The next thing to do is to personalize your account. Enter all the details asked for and once your account is active, you will need to put up a profile picture. Obviously, for your personal account, you can set up whatever you'd like. For your business account, however, the best idea is to put up a picture of your company – you could use a picture of your trademark or logo or even a picture of your office space with the headboard of your company's name on it! remember, your picture is going to speak to your audience about what your business really means, so keep it simple, elegant and let it speak for itself.

Although there is limited space to add company details, the short profile bio can be used to share a link to your company website, so be sure to include an accurate web address that users can be redirected to. Brand and campaign hashtags can also be used in the bio to encourage people to search and adopt them in their own posts.

Instagram will offer you the option of entering your bio on your profile – this is basic information that anyone can see when they get to your page. You can change your settings to make your entire profile private; obviously, as a business account, you want to leave it public so that everybody can view it. In the bio, provide details about your business – do not use long sentences, especially since you have only a total of 150 characters! Make it simple, fun and catchy so that they'll remember you even after they leave your page. You can also leave a website link in the bio so that they can log into your official website and get more details from there.

Use a high resolution image for your profile picture, preferably a logo or team photograph, as this will be a representative of your brand.

Now before you start using Instagram to promote your business, I would suggest you experiment with the app itself and learn more about how it works. This way, you will be able to make full use of it – create a few personal accounts and then start exploring the various options it offers to learn the ins and outs of it. for instance, explore your settings to see how you can get your notifications to appear – internally or externally, even after exiting the app. Learn how to manage your feed, how to use the Direct Messaging feature, how to post and edit pictures, how to Follow and Unfollow people, comment on a picture, moderate comments on your pictures, how to Like pictures, how to tag photos, etc., etc.

Before you start following other profiles, make sure that you have around 5 good quality images or videos uploaded, as existing content will give your profile credibility and optimise your chances of a follow back. When you view an Instagram profile from the web, a mosaic of randomly-chosen Instagram photos that you have uploaded will appear on the top of your profile.

Once you know how to do all that, you can get around to making Instagram an official part of your marketing strategy! Here is another thing to keep in mind – connect your Instagram with all other social platforms as well! As I have mentioned previously, the power of social media makes a huge difference to a small business if used correctly. This means that having an Instagram account alone cannot be the ultimate end for your company – you need to promote yourself via Facebook, Twitter and whatever other platforms you use to speak with your customers! Fortunately, Instagram offers you the option of linking everything together – in your settings, you will be able to choose the network you want to sync your account with. So this way, anything you post on Instagram will be accessible through the other sites too. Again, on the other websites also, stick to the same username so that it's easy for people to identify you and your brand.

Now you are ready to start posting on your Instagram to promote your business!

Making Your Account Look Attractive to Get More Followers

Now that you have your account on Instagram set up and ready to go, you can start using it to promote yourself! In this chapter, we will look at how you can make your profile look attractive and well made so that your customers like the look of it; these ideas are simple tips that you can follow to make both your private as well as your business account well liked! Here's what you can do to make your account more visible -

- Make sure all your posts have a recurrent theme! Now, as a business venture promotion, this is actually a very easy thing to do – you can start by posting pictures of your products or maybe of customers using your products. Let me give you an example – we will be using the idea of the Awesome Girl Apparel throughout the rest of the book. For instance, let's assume that you are launching a dress of new design. So for the full length of that campaign, you should post pictures of the dress, preferably with models wearing them to show off how it looks. You could post different colors of the dress, different sizes, etc., - the goal is to make the viewers understand what the features are through pictures instead of words. This kind of a theme will let your customers and Followers know what to expect from your account.

- Make sure the pictures that you put up are of good quality; there's no point putting up something meant to promote your product if the viewers can't even see what it is properly! And don't just keep posting simple camera clicks – use the editing options that are available on Instagram or use more professional editing software like Photoshop on your computer to make the photos interactive and gorgeous. Stick to your theme, keep it elegant and professional and show off what you have to offer.

- When you are just beginning your account and these are going to be your first few posts, then make sure you start out with a splash! If you already have a Facebook page or a Twitter Account, then chances are that a lot of your followers from there are going to Follow you here too; make your profile interactive and interesting. Your first few pictures and posts should be well planned, well thought out and strategic; I would advise that you start with your theme right away! It doesn't have to be anything special; it can be as simple as the products you've sold this week or the sale you'll be having next month. Make it fun, make it interesting so that you'll have repeat customers who will Follow you!

- A picture is worth a thousand words, but Instagram does give you the option of adding a few words of your own as well! Caption your picture well – remember, you have only a few characters to say whatever you want to say, so keep it short, simple and catchy! For instance, let's take the case of that new dress Awesome Girl Apparel is introducing. If we put up a photo of a model wearing that dress, you can probably write the slogan for that campaign on it, something like – *For the Awesome Girl in you...* you could also add the prices, key features and other ideas, though I would advise you to stay away from long descriptions like that. In such cases, it's best to leave a link to your website where they can find details – this is just a glimpse of what's in store for them!

- Link every other social media account that you have to your Instagram Account. Whether it's Facebook, Twitter, Tumblr or any other blog you run, it's essential for a business owner to be accessible to their customers at all times! Your bio must have a link to your company website so that interested viewers can click on it to get more details about you. If possible, add your professional contact number or mail, so that they can speak to you personally. But keep in mind, your Instagram account must be slightly different from your Facebook page or your Twitter Feed. Keep your posts separate but linked with a common theme; that way, your customers will need to Follow you on all platforms and not just one. And as the prime rule of business goes, repetition is the best way of convincing prospective consumers, so showing up on their feed constantly, no matter which social media platform, is going to prove beneficial to you.

- The most important way to get noticed on Instagram (or any of the other media platforms as of today) is to hashtag. Hashtagging has become essential and vital if you want more views for your posts; whether it's Instagram, Tumblr or Twitter, you need to tag your posts to whatever theme you're following so that people can easily find your posts. The tags are what they use to search for pictures or posts. You want your tag to be one of the ones that are trending; depending on how well it's trending, the number of followers to your page will go up! For instance, if Awesome Girl Apparel is introducing a new Plus Size wardrobe, then it's not enough to stop with tags like *#AwesomeGirl* or *#PlusSize*. You can tag related ideas like *#bodypositivity* or *#everygirlisanAwesomeGirl* or *#beautybeyondsize*, etc. Thus, someone who's searching for posts on body positivity will stumble upon your blog and maybe even end up becoming your customer!

- Interact with your customers! This is especially important when your account is at its fledgling state – the best way to get more Followers is to be extremely active by talking to your customers and your Followers. Obviously, since it's a business account, you'll need to be careful about whom you're Following in return or what you speak to them about, but don't let that stop you from interacting with them at all! In fact, a lot of customers are happy to get the personal attention from the business owner itself – it makes them feel important and they will definitely pass on messages about you to their friends. Word of mouth marketing will do wonders at this point for you! Encourage interaction via comments and Direct Messaging.

- Hold regular Instagram contests to engage further with your Followers. This is something all Instagrammers do regularly, even on their private accounts. I've expanded more on this later, along with tips on how to host a successful contest.

- Space out your posts properly. Yes, you want to be a recurrent appearance on your Followers'

Feeds, but too many posts and it's possible that they'll get bored or annoyed and end up Unfollowing you instead, which is the opposite of what you want. This is especially true when they're following you on multiple social media networks; for any marketing strategy to be successful, you'll need to draw a fine balance between pushing your idea forward and thrusting it down their throat! So time it well; for instance, Awesome Girl Apparel will stick to one post per one social platform per day. Today's Twitter Feed will show one model in the new clothes, while tomorrow's plan is to post another size and another color of the same dress with a different model on Instagram.

- Ask your Followers what they'd like to see! As I mentioned before, interacting with your customers is the best way to gain even more Followers; you should have regular Question and Answer Sessions with them on both Instagram and other platforms to ask them what they would like to see. It's possible that your prospective customers from the other side of the world want to see more close up shots of your product, or find out something about your production process – unless you interact with them and get their feedback, you will never know.

- Another important way of tagging is to Geotag. As the name suggests, this type of tagging is basically letting your customers know the location of the photograph; when you tag the location of your photo, people searching for pictures of that photo will be directed to your profile and page and get to know about you! For instance, let us say that Awesome Girl Apparel had an outdoor shoot for the promotion of a swimsuit campaign in a place like Miami Beach. Then, if I added tags about Miami Beach, then it's possible that someone looking for pictures under that tag will become my prospective customer! Tagging the locations also adds a very personal touch to your posts that lets your Followers know what you are up to and what is happening with your company – it helps to build brand loyalty in a very visceral way.

Since bursting onto the scene in 2010, Instagram has exponentially grown in the last few years, and is now one of the most widely used social networks with more than 100 million monthly active users.

With such a wide audience of active users, everyone from small local business owners to major multi-national brands are finding creative ways to use Instagram as part of their marketing efforts.

If you are just starting out on Instagram, or just starting to use Instagram as a marketing channel, here are some tips on how you can promote your brand/business, drive awareness, and hopefully convert customers and drive revenue.

How to post on Instagram

How to take photos and videos?

When launching the app, a blue camera icon will appear at the bottom of your screen and from there, you can select either photo or video mode. Take your photo or record your video, as you would normally via your mobile camera and click the 'next' button when you are ready to proceed to editing.

You will then be given the option to crop the photo/video as you please. On the next page there are a number of filters that you can add to your content - pick the filter that looks best! You can also upload your content without any filter by using the default option: 'Normal'. If you would like a little more control over your editing, you can find contrast and settings icons on the toolbar that can be used to adjust the contrast, shadows, saturation and more. Once you're happy with your image/video, click on 'Next' at the top right corner.

At the top of the new window, you can choose between sending this photo to all of your followers or specific users that you can pick on a one-by-one basis. Add a caption to the photo and add a link to your website when appropriate. Tag people whenever you can, they will receive a notification and are likely to engage with your post. You can also add the location - users searching this particular location will then see your photos.

If you want to share this post on other social networks such as Facebook and Twitter, just select the button of your choice.

Once everything is ready to go, click on 'Share'.

1) Getting Started

The first thing you'll want to do is set up your Instagram account. Pick an easily identifiable username (I would recommend just using the full name of your business/brand), write a short description (preferably with relevant keywords you are using elsewhere in your digital efforts), and add a link to your company website.

Another tip that can help your profile get found is by using hashtags in your profile description. For example, if you are a ski apparel brand, you could use hashtags like #ski #snowboard or #fashion in your profile description.

2) Follow and Engage with Existing Relevant Communities

If you've done any marketing on other social media networks, you are probably already well aware that engaging with your community is a great and almost instant way to increase your exposure, gain followers and gather important marketing insights.

Spend some time browsing Instagram to see what people not only in your industry are posting, but also what your customers and competitors are posting. See what photos they engage with, who they follow, and which photos they like and comment on.

From there, you can find people you consider might be your key demographic and start to engage with them. Like and comment on their photos. If they are interested in what you do (or what your company offers), they will be likely to engage back and follow your account.

How to enrich and nurture your community

The center estimation of Instagram is its engagement rates. To effectively draw in, you have to contribute. Offer photographs of your group, occasions you have or are included in and substance, for example, in the background at work. This is your opportunity to advance your image as a respectable and relatable association.

At the point when sharing substance, ensure that it is relatable for your devotees in the meantime as being pertinent to your association. Consider what your objective business sector are keen on, and afterward fuse those subjects into your Instagram content in a way that can identify with your image.

Posting recordings is additionally certainly justified regardless of your time as they can produce 3x the measure of inbound connections than composed posts. Instagram have as of late taken a leaf out of Vine's book and presented an auto play and circling highlight, so recordings will begin naturally and keep on playing until the client effectively stops it, which means more video perspectives.

Likewise with other interpersonal organizations, facilitating giveaways and challenges will help you to collect more engagement and increase new adherents. At the point when dispatching a battle, develop a hashtag that can be effortlessly received by other Instagram clients, as it will urge individuals to take an interest and will fabricate a group around that hashtag.

Joining forces your image with a decent motivation is another way that you can pick up a bigger after for your organization, as a preformed group will have as of now been made in backing of that bring about. This could be anything from gathering pledges for a philanthropy, to supporting the estimations of a battle, for example, reasonable exchange or solid living.

3) Mention Influential People and Brands in Your Posts

I recently helped a heli ski operation in Alaska, Alaska Snowboard Guides (ASG), get their Instagram account up and running. As part of this effort, one of the first things I did was search Instagram for clients of the operation who had just flown with ASG, and then follow the ones who had accounts.

- **Comment on other photos with hashtags**

If there is a hashtag you are regularly using on your business Instagram hashtag, it can be wise to spend time commenting on relevant photos from other Instagram users with that same hashtag. Make interesting, relevant and non-spammy comments and include hashtags where appropriate.

Grant Kennedy

This will make it easier for users that are looking for pictures or people with these keywords to find what they are looking for.

- **Create or use a hashtag for an event**

If you are attending a conference or event (or hosting one) a great way to gain additional exposure for your photos and Instagram account is by either creating or using an existing hashtag for that event.

For example, if you are an outdoor industry company attending a major tradeshow like "Snowsports Industry America," tag any photo updates from that event with the #SIA13 hashtag.

5) Don't Forget to Promote Your Instagram Presence

This seems fairly obvious, but a great way to get more exposure on your Instagram account is to make sure you are promoting the fact that you are on Instagram and what your handle is, so people can find you.

Add This, a company that makes social plugins for brands and publishers, has an Instagram follow button for websites you can easily add in minutes. You can also write a blog post about your Instagram presence, tweet out the news, post to Facebook, create a pin on Pinterest and more to get out the news that your brand or company is now Instagramming away.

Chapter 3

Why Instagram is important for your
company or business?

Branding: The goal of branding is to create an emotional connection between a company and its consumers. Branding results from the sum of many different parts, including the brand name, logo, colors, and more. It allows companies to differentiate themselves in an increasingly crowded market. The rise of new technology is changing how companies communicate with consumers and in turn how they use branding.

Instagram is the perfect tool for showcasing your company's values, culture and staff. Behind the scenes insights of your company at work can offer a feeling of exclusivity and help users to learn more about your company and relate to your brand. Be sure to be consistent with the quality of content shared to maintain a positive representation for your brand and to keep hold of followers.

Interaction: Engagement on Instagram is higher than elsewhere - it's a great platform to interact with your community. By posting images that provoke a response, you can build a relationship with your followers. Encouraging user generated content can be even more effective than posting content of your own. By incorporating a hashtag into a campaign, you allow users to get involved themselves, at the same time spreading your message and later in search. You can encourage users to participate and award users for taking part by sharing their contribution on your profile or website, or by offering a prize for the best entry.

Community: Instagram communities are really active, remember that most of its users check their newsfeed at least once a day. A community can be built by incorporating actionable hashtags into your content or developing campaigns around an upcoming occasion or relatable theme. As previously mentioned, encouraging your followers to join in with campaigns by providing their own content can help to build a community around your brand.

Chapter 4

Using Instagram for Business – The Basics

Instagram, with more than 40 billion images shared and 400 million monthly active users, generates an average of 80 million photos per day. The mobile-based photo- and video-sharing social network powers the sharing of images and creation of community among users around the world. At only six years old, the platform has shown significant growth in its overall user base and in almost every demographic group.

As people join Instagram in droves, brands have a unique opportunity for engagement with their fans: Instagram posts generate a per-follower engagement rate of 4.21%, which is 58 times more engagement per follower than Facebook and 120 times more than Twitter.

Success for brands on Instagram takes more than publishing attractive images—it is the product of thoughtful strategy, a well-defined brand identity grounded in visual creativity and effective community management. As you explore the potential of Instagram for your business, keep in mind the particular strengths of visual media for telling a compelling story about your brand.

When you put these principles into practice and combine storytelling with stunning images—like Flowers for Dreams, which was recently featured by Instagram for Business—your brand will reap the benefits. To help you develop an Instagram marketing strategy based on clear goals and measurable results, we've put together this comprehensive guide.

Determine Your Objectives

Whether you haven't distributed a solitary photograph or you need to lift your current nearness, consider the accompanying while making your methodology for Instagram:

•   What will Instagram permit you to do that different platforms don't?

•   Who is your intended interest group and which individuals from your gathering of people are dynamic on Instagram?

•   How will Instagram incorporate with alternate systems in your online networking methodology?

Instagram's emphasis on visual sharing offers a one of a kind stage to showcase your way of life and individuals notwithstanding your items and administrations. The versatile way of the application

Grant Kennedy

fits rapidly catching minutes, allowing supporters to connect with your image in a way that can feel more easygoing and prompt than on different systems. Contingent upon your industry, image and key execution markers, your Instagram methodology may focus on a few of the accompanying goals:

• Increase brand mindfulness.

• Demonstrate organization society.

• Showcase your group and enroll new ability.

• Increase client engagement and devotion.

• Showcase items and administrations.

• Enhance and supplement occasion encounters.

• Incentivize customer engagement with your image.

• Share organization news.

• Grow your group.

• Connect with influencers.

• Drive deals through an outsider application.

As you keep on developing your system, these targets will control you in deciding the best way to deal with every part of the procedure.

Develop a Content Strategy

Content is the foundation of your Instagram presence. Many businesses use Instagram to make their product the star of the show, while many companies often focus on company culture and team recruitment—the right approach is one that best showcases your brand. Based on your target audience and objectives, develop a plan to deliver eye-catching content to your community on a consistent basis.

INSTAGRAM + CONTENT

Creating content on Instagram Content is central to the Instagram experience. It's why tens of millions of people visit the app every day--to view beautiful, interesting imagery in their feed and to post their own unique photos and videos. The sharing of this visual art is what makes the Instagram community so dynamic and engaged. Your brand's content should add to the experience of being on Instagram for your followers. We recommend the following guidelines for creating content on Instagram:

1. Identity & voice: Develop a framework for bringing your brand's identity to life on Instagram, based on your business objectives. Identify words that reflect your brand's voice and tone; the feelings you want followers to associate with your brand; and the role you want your brand to play in their lives. This framework will inform your content, and in turn the

experience that followers have when viewing your images.

2. Content themes: Establish regular content themes, or pillars, that are authentic to your brand and fit the Instagram platform. Ensure that your posts adhere to these pillars. This allows for a diversity of content that also remains consistent over time. Followers will know what to expect from your brand on Instagram as you reinforce key brand associations.

3. Image subjects: Post photos and videos of unexpected and behindthe-scenes moments that feel authentic and immediate. Candid, insider access is what people love about Instagram. We recommend avoiding overly promotional images or those that are simply repurposed from other channels. These images appear out of place on the platform and detract from establishing a clear and differentiated brand identity and voice. 4.

Image enhancement: Adjust your images with filters and other tools available through the Instagram app. These effects give images that unmistakable "Instagram" look that people respond to.

5. Text: Keep captions short and fresh. Incorporate hashtags where relevant, but not so many that they detract from the simplicity of the post. (Refer to section on hashtags for more guidance). Ask questions in the captions of your images to engage with followers.

6. Location & people: Include the location of your photo or video when it helps tell the story of the image

(i.e. it was taken at an event, roadshow, retail location, company headquarters). Use the Add People feature to tag accounts in your image when they will help you reach a broader audience and you have permission (i.e. partner brands, celebrity spokespeople, etc.)

7. Timing: Moderate the number of posts you make per day to ensure a consistent but non-intrusive presence. We recommend anywhere from 1-3 posts per day. Experiment with posting at different times of day by monitoring engagement. If posting content from a live event, consider creating a separate account to avoid annoying your followers by taking over their feeds.

Creating video content on Instagram:

Instagram recently launched the ability to capture and share short videos on the platform. While businesses are still experimenting with this new feature, we recommend the following guidelines:

1. Include video as part of your content strategy consistent with your identity, voice, and content pillars.

2. Utilize features outlined above when they enhance the quality of the content, such as filters and hashtags.

3. Consider which unique moments in the life of your brand are best shared through a moving image. Here are a few examples:

• Share an inside look at how a product gets made, or how new products are developed

- Preview a new product or service and show how it works
- Offer short product tutorials
- Show how your brand fits in to people's everyday lives
- Capture moments from brand events or experiences

Making content discoverable on Instagram:

Hashtags, Photos of You, Photo Maps Instagram offers useful tools for you to find content posted by other accounts about your brand, and for people to discover the content you post. You can tag photos with topic (hashtag), account (Add People), and location (Name This Location). We recommend the following:

1. Hashtags: On Instagram hashtags serve many purposes for brands. Use them to classify the images you post, for instance tying them to different content pillars. Use them to reach new people who may be searching a hashtag related to your brand. Or use them to support campaigns or contests aimed at driving awareness by asking people to tag their photos with a hashtag associated specifically with your brand. Hashtags should be used in moderation and when relevant, not simply to capitalize on a current event or popular hashtag.

2. Add People (Photos of You): While hashtags are a useful tool, we also recommend encouraging your followers and brand advocates to add your official account to their posts using the Add People feature, so that they appear in your Photos of You. Photos of You can be accessed from your Instagram profile whereas

hashtagged images cannot. Hashtag images must be searched for and often surface images with no association to your brand (for example, consider the hashtags #apple or #target). Many businesses set up manual approval for Photos of You so that they can approve content before it becomes part of their profile.

3. Name This Location (Photo Maps): You can add a specific location when posting your content on Instagram, which adds the photo or video to your Photo Map. This feature is most useful when posting content from an event, roadshow, retail location or company headquarters. Your Photo Map can be accessed from your Instagram profile and gives users a new way to explore the images and videos you've shared. When you add location to an image, the image can also be seen by anyone who views that location in the future (by clicking the location on a post).

Build Content Themes

Review your objectives and determine what aspects of your brand to showcase in your Instagram content. Products, services, team members and culture all offer rich potential for subject matter over time. Once you have a list of specific content themes, brainstorm possible subjects for your images and videos.

Some companies focus on showcasing their products and services, offering practical tutorials or going the opposite direction and creating whimsical tableaux with their product as the hero. Dunkin Donuts' colorful feed puts their offerings front and center, showing off both standard and seasonal offerings by tying post content to major holidays and events.

Determine Types of Content & Ratio

Instagram started as a photo-sharing app, but its wide base of creative users publish everything from videos to graphics to animated GIFs. As you plan out your content, consider a balance of content types that will work best for the resources you have and the engagement you want from your audience.

If video enables you to tell a compelling story about your product, work it into your content more often. If you don't have the resources—time, skills or comfort level—to execute video at the level you want, you may choose not to publish video at all or to reserve it for specific campaigns and promotions. With composition for Instagram, quality matters, and it is worth spending the time to create the best possible content.

Beyond its flagship app, Instagram offers several supplementary apps that help you get even more creative with your posts. The Instagram suite of apps includes Hyperlapse, Layout and Boomerang, which empower users to create time-lapse video, image collages and GIFs, respectively. These additional apps allow brands and consumers alike to create even more unique, Instagram-specific content even without in-house design or video production capabilities.

Set a Content Calendar, But Be Flexible

To establish and maintain an active presence on Instagram, determine the frequency with which you will post. Then you should develop a content calendar that cycles through your themes and integrates key dates and campaigns. Instagram does not have a scheduling function, and it does not grant third-parties API access to publishing, which means you cannot schedule posts directly on Instagram or through your social media management tool. That said, you can easily prepare content (photos, videos, captions) in advance and create a content calendar so that your team knows when posts should go live.

Some of the best content for Instagram will occur spontaneously, especially if your aim is to highlight company culture or events. By preparing content and setting a general schedule in advance, you can allow the flexibility to take advantage of opportunities when they occur. During events, be ready to publish quickly to take advantage of real-time social engagement.

Consider Curating User-Generated Content

If your Instagram community members are sharing their own content featuring your brand, you have access to a repository of potential content gold. Curating content from your fans allows you to foster audience engagement and create an incentive for your audience to share their own creative ways of interacting with your products, services or company.

As always, the photos you choose to curate should match your brand aesthetic. Make sure to review users' accounts and other posts before sharing their content in order to judge whether it is appropriate to publicly align your brand with them by sharing their photo. In terms of best practices, it's courteous to ask someone for permission before sharing ("regramming") a photo. Always give credit by @mentioning the original photographer in your caption, and provide your fans with insight on how to share more photos that your brand might feature in the future.

You can find user-generated content on Instagram by monitoring your branded hashtags and business' locations. Sprout Social's Instagram monitoring tools make it easy to set up brand keywords for Instagram hashtags and locations so that you can both monitor and engage with fans' posts in the Smart Inbox.

Pottery Barn does an excellent job of increasing brand awareness by encouraging Instagrammers to use the hashtag #mypotterybarn in order to have their photos shared by the brand. When regramming a customer's photo, Pottery Barn uses the caption to comment on the original, often giving a compliment to the photographer and always referring to the specific Pottery Barn item(s) featured.

What started out as a place to socialize with people and widen your friends' circles has today become a platform for advertising and marketing your product. Given that it is a worldwide presence, small business owner now has access to a much larger customer based than earlier on; it is advertising that can be done at an extremely low cost, especially in comparison to the previous decades when one had to spend thousands of dollars to get an ad displayed on the TV or played on the radio.

The app itself is not a service that provides products; as you can see, it is simply a tool for you to make use of within the limits of your overall marketing strategy. Over the few years that it has existed, the app has gained a huge amount of popularity, especially amongst businessmen, so much so that the owners decided to make it a free service to be provided to them. A blog called Instagram for Business was begun – you can log into that website and learn how to make use of the app to augment your business strategy.

The blog also provides a platform for you to connect with other businessmen using Instagram for marketing their own ventures. It gives you detailed guidelines on what kind of simple advertisements you can create and post on your account, how you can partner up with other businessmen and what all you can do to use this mobile photography platform to promote your small venture and build brand loyalty amongst your customers.

Chapter 5

Making Your Posts Profitable

Here are a few simple things you can do to make sure that your posts allow you to make a profit –

- Make sure your photos are appealing, well edited and provide a good idea of what the product actually is. For instance, in the case of a clothing store like Awesome Girl Apparel, sizes can be a problem, especially for the extremely big or extremely small sizes. This means that we will need to put up pictures of models of all sizes and reassure our customers that they will not be cheated if they make a purchase online without physically coming to the store.
- Provide a link in either the caption or the comments section to the website where the customers can make their purchase or get more details. If, for instance, we upload a picture of a particular type of dress, available for a limited period of time, then we must upload the link to the site where they can make their purchase. It makes it easier for the customers to gain access.

If you don't provide the link, they may get frustrated searching for it and let go of the purchase, even if they are interested and you've lost a prospective customer.

- Don't make your posts overly crude; as I mentioned earlier, you want to toe the line between marketing and thrusting your product upon your customer. Your product quality will speak for itself – what you need to do is to convince them that you are worth trying out in the first place. So keep your posts simple and let them speak for themselves. Don't make your posts a catalogue alone; it's not just your product, but your brand you want to market. That means you'll need to include your employees, your investors, you shareholders and any of the other stakeholders of your business in your posts to make it that much more expansive.

- Keep your Direct Messaging feature open at all times to interact directly with interested customers. Sometimes, they may prefer to speak to you in private instead of the public comments section; you will have to be open to interacting with them as such to make sure you collect more customers! Send them video links, pictures or any details that they have requested and also collect feedback from them through this option.

- Have regular giveaways and contests on Instagram. I've explained more on how these work in the following chapters; when you host these regularly, you will gain a lot more visibility and traffic to your profile.

Gaining More Visibility Via Facebook

As I have already mentioned, you must connect all your social media accounts together so that you are easily accessible to your customers at all times. You can make use of one account to promote the other; for instance, if you're hosting a Giveaway Contest on Instagram, use your Facebook page to promote the contest.

Facebook as such has a larger user interface and traffic than Instagram; if you want to generate more interest and likes for your posts, link your accounts together! You can also post your Instagram advertisements on Facebook – it is easier for your Facebook fans to share your ads on Facebook than on Instagram. You can also run a campaign that gives extra incentive to all those who follow you on both accounts – your objective is to promote your brand, so come up with new and creative ideas that you can use!

Gaining More Visibility Via Your Official Website

You can promote your Instagram page on your official website too; it goes without saying that this is the site that gets the most number of viewers out of all your accounts. Direct this incoming traffic to your social media accounts so that the customers become repeat customers and form a part of your loyal network. On your official website, the viewership is generally just one view per person; if they Follow or Like you on other sites, however, you can keep them up to date with your business activities and make them part of your network and brand. To do that, you need to encourage them to visit your social media sites, so use your website to promote your Instagram, your Facebook and your Twitter so that they can visit these places and Follow you.

Gaining More Visibility Via Hashtagging

I've already mentioned this earlier, but that I am repeating it again should tell you how important a step this is. The trending hashtag is the most effective method of directing traffic to your profile, so make full use of it! Use the Explore page on Instagram to identify what tags are trending – these are the tags that people are searching for the most. See how you can connect these tags to your business – make sure they are legitimate reasons to do so and not some flaky excuse that will leave prospective customers bored! Analyze the trending tags and learn to use them in full.

How to Use Instagram Direct

In late 2013, Instagram introduced a new feature: Instagram Direct. One thing that Instagram lacked was the ability to maintain a private conversation with any one person or small group on Instagram. Now, with the introduction of Instagram Direct, we can have conversations on a more private level.

You will notice the little inbox icon in the top right corner of your home screen – this is the only place to access the inbox. Notifications will appear here when you receive new messages.

You can send direct messages to one person or a group of up to 15 people. You cannot send text only messages. You must initiate a direct message via a photo or video post. Comments can continue on the post as text based conversation after the message is initiated.

When you are in your inbox, you can see the messages in which you've participated. Any message with your name is a message that you initiated. If the message has someone else's name, they initiated the message.

There are two ways to generate a new message: through the inbox directly or through the regular upload process.

When you're in the Instagram Direct inbox, you can click on the + symbol to generate a new message. Take a new photo or video or upload one from your gallery. Edit your post with filters as you normally would.

If you don't use the Instagram Direct inbox, you can upload a photo or video from the in-app camera or your gallery. Edit your post with filters as you normally would.

You'll notice that when you get to the caption stage, there are two tabs at the top of your screen: One for Followers and one for Direct. If you are going to send a post via Instagram Direct, you'll want to make sure that the Direct tab is highlighted. If you decided you didn't want to send this post privately, just click on the Followers tab and the post will proceed as a normal post upload.

Add a caption and select the people to send the message to. Instagram will suggest users or you can begin typing a name in the search bar. When the name of the person appears in the list, click on their name and a green check mark will appear in their circle. Their image will also appear in the "To" field. You can select up to 15 people. Once complete, click the check mark in the top corner or "Send to (#)" button.

If you want to respond to an original conversation with another photo or video, you cannot add these to the conversation stream. You would have to start another message with your photo or video. This is one thing that I expect Instagram to improve upon and update in the future.

When viewing messages, the person who started the message will appear on the left. Everyone who was sent the message will have their image appear in the "To" field. Anyone who has not seen the message will remain faded. Anyone who has viewed the message will appear with a green check mark. And anyone who has viewed and commented on the message will appear with a blue chat icon.

Comments will appear beneath the post and profile pics of those in the conversation. Any comments left on the post are visible to everyone invited to the message. But only those in the message can view the comments and the post.

Technically, anyone can send you a direct message. Or, you can send a message to anyone.

If you want to send a message to someone you don't follow, you need to know their username and type that into the search field. Their username and profile photo will appear and you can add them to the list of recipients.

However, if someone sends a message to someone they do not follow, the recipient will receive a message that this person is trying to send them a message. The sender's name and profile photo will appear with the option to approve or decline the message. If you accept to receive this message from this sender, they will be able to send you messages again without your approval.

This is a good method for Instagram to control spam and messages from unknown users. However, be careful who you accept messages from. You don't see a preview of the post prior to accepting and if you allow them in once, you automatically allow them in again.

You can choose to block this user, however, if their messages become unwelcome in the future.

You can delete messages you have sent or delete certain comments from the thread of the conversation on a message you started. Clicking on the triple dot button or edit button will open a pop-up menu with options. Likewise, you can delete comments you made on another person's message or hide their post using this menu.

So, for those instances where you would like to have a conversation in a more private environment, Instagram Direct allows you to manage direct messages.

Chapter 6

Hosting an Instagram Contest or Giveaway

As I mentioned earlier on, holding a contest on Instagram is a brilliant way of getting more Followers as well as promoting your business venture. It provides an excellent way of generating interest from prospective as well as existing customers and it also allows you to have personal, one on one interaction with them! Materialistic as it sounds, the prospect of a reward at the end of the contest is enough to get a large number of people to Follow you – once they Follow you, it's up to you and your posts to convince them to continue to Follow you. Giveaways are just another word for contests.

Choosing what you are going to give away as a prize for your contest winners is a very important decision to make, especially for a business account. You want new customers as well as repeat customers. Obviously, you will choose some part of your product to give away as the prize, but you must make sure that it's of good quality so that the winner does not feel cheated and promotes you through word of mouth, both physically as well as on social media. For instance, at Awesome Girl Apparel, we would probably use this as an opportunity to give away an extra piece of a design that would otherwise be sold out in our retail outlets. This way, contests allow us to perform clearance sales of a sort, but remember – customer satisfaction comes first!

Another idea is to get a custom made prize for the sake of the contest in particular, but this is a more expensive option, since you will need to spend a good amount of money on it. You could also allow your winner to pick out a gift directly for themselves from the range of products you offer or you could give away a free gift card worth for a select amount of money for which they can buy whatever they want. Again, these are expensive options, so make sure that your budget covers all of it before you jump head first into it.

Also keep in mind that winner announcements are important posts; you don't want to make it too ostentatious so that the other customers don't feel left out, but you don't want to not make a big deal out of it either. You also want follow up after the contest is over to make sure that your customers are enjoying their gifts. Checking up on them will ensure repeat customers and posting their feedback on your account will allow you to augment your overall advertising. For instance, if the winner of the Awesome Girl Apparel contest received a new designer bag as their gift, we would probably make an announcement with his or her picture on our profile. Maybe a day or two later, we would request that they allow us to post a picture of them using that bag in public, with a singular, catchy caption about how good a product it is – that way, the winner knows we care for their response and we get to promote our brand further.

How often you will hold your contest depends on a number of factors. Hold a trial run first; depending on how much it costs, how much interest it generates, how much you will invest and how much visibility you get, decide how often you'd like to do it. Sometimes, it may require a lot of effort and time and sometimes, it may be as easy a single post – it depends on the contest, so be open and flexible.

Now, even the kind of contest you will hold isn't set in stone; remember, each business differs from the other and what suits one type of venture may not be suitable for another. It is the same with individual accounts too – your personality differs from someone else, so what they do may not be applicable to you. So you will need to customize the kind of contest you are going to conduct for your specific needs; however, I have listed below some of the most popular and basic kinds of contests that are often conducted on Instagram as a starting point for you. Read through them and then see what works for you – rework it according to your business model so that it will fit your venture's overall marketing strategy.

Contest #1 – Random Selection Contest

This is probably the simplest and the easiest way of holding a contest on Instagram. The basic premise is that you put up a post asking your Followers to like a particular photo or Follow a particular post. Give them a specific deadline within which they must do so; that is the length of the contest's validity. Once the contest is over, you will go through the list of all the people who have liked your post and then, at random, select your winner and then give them the chosen gift. Again, ask them to post a picture with the prize as feedback and give you what is known as a shout-out in their comments section.

Contest #2 – Maximum Likes Contest

As the name suggests, you will choose the winner of this type of contest based on who gets the maximum likes for their pictures. What the participants have to do is simple – post a picture with a specific hashtag that you have created for this purpose or following the theme or the guideline you have set for this particular contest. Now, since the objective is to promote your business venture, a good idea is to ask your customers to post a picture of them using your product; remind them to tag your Instagram account to their picture! Again, set a deadline for the contest's validity and the winner will be the person with the most number of likes on their uploaded photograph.

One of the main reasons why this kind of contest works is because people love to post selfies! It's easy and they have nothing specific to do as such. For instance, if Awesome Girl Apparel were to do this contest, we would ask our followers to post a picture of them wearing any of our clothes and/or our accessories; the picture should emphasize the product and tell us (without words, obviously), how much they love using it. Since people love to get dressed up, take pictures and post selfies, this type of a contest, especially for a small boutique, will provide a huge amount of visibility for us!

Contest #3 – Walk into the Store Contest

This is an excellent method of increasing traffic to not just your social media accounts, but your actual store as well! The rules of the contest are simple, if a bit demanding of your customers – they will have to walk into the physical store and then click pictures of themselves within the store, which they will then upload on Instagram and tag your account to it. Now, the problem with this kind of contest is that it works with a very specific type of product; for Awesome Girl Apparel, it is a great idea! They can try on new clothes, post the pictures online and then we would select the winners. And as a bonus, the prize they receive could be the very dress they posed in.

But for a different product, like a bookstore or an electronics store, it may not work out so well. You also want to make sure that the picture they take is used only for the purpose of entering the contest and not for anything else; at Awesome Girl Apparel, they could simply take the picture in the new dress they're trying out but not upload it for the contest and instead use it for personal purposes.

Despite these hiccups, this is an excellent method of directing more physical traffic to your real life facility instead of just your virtual shop! Again, remember to specify or create a special hashtag for the contest, set up a contest validity deadline and announce the winner with the follow up feedback after the contest is over.

Contest #4 – Website Contest

This is for the more elaborate kind of contests – the actual challenge itself does not take place on Instagram, as all the previous challenges have been. Instead, you make use of your Instagram account to promote an elaborate, more difficult contest that you are conducting through your company's official website. This means that instead of putting up posts with your customers' pictures, you will need to put up regular reminders and contest details.

This is where your editing skills come into play; you will have to make your photos give details about the contest and provide the link to your website in your captions. Make your pictures fun, colorful and easy to like – they should also be in keeping with the theme of the contest! Put them up every so often to remind the followers that the contest is happening, just on a different platform and that they will have to log into the official website to take part! You're going to have to be extra convincing, since it's extra work for them – make it attractive with details about the prize and what's in store for them if they do take part.

Contest #5 – Video Contest

Mobile photography includes both photographs as well as videos and Instagram allows you to upload both! However, videos on Instagram have a limited length to adhere to. So keeping that in mind, a video contest is a good idea to get your customers more involved with your product. What they will have to do is post a short video, of maybe ten seconds or so, wherein they are using the product. The caption can be why they like it or what they find lacking in it – the winner can be someone of your selection or the person whose post has the most likes. It's your choice. For instance, at Awesome Girl Apparel, we would probably ask them to post of video of themselves in a full outfit that they have designed themselves using our clothing and/or accessories.

This type of contest generates a lot of interest because, once again, people love posting pictures of themselves. They would also want to make themselves look good to their own followers, who will see everything they post, so your product will get a good, wide range of visibility overall. Again, don't forget winner announcements, follow up and feedback and the prize itself.

Contest #6 – Shout-out Contest

This is a very simple way of getting more publicity from your followers. As you probably know, on Instagram, one way to get more followers is to do shout-outs; this basically means that when you post something, you tag another account to it and give them a shout-out, asking your followers to go and check out that person's account and if they like it, Follow them too. This kind of contest follows that premise; the winner of the contest will be the person who has given you the most number of shout-outs within the contest's specified deadline. Of course, since it's not a very big or difficult thing to give someone a shout-out, the prize will have to be something that's not too expensive for you; you could use this as a good clearance sale option or have the winner thanked publicly on your website.

These are just some of the simplest ways in which you could conduct contests on Instagram to promote your product and your business venture. Every person loves to receive free gifts; by offering your prospective customers a little something in return for participating, you manage to garner yourself a loyal customer base that will return to you and your venture. Their word of mouth promotion also makes a huge difference – remember, when someone takes part in your contest and post about it on their account, *their* followers get to see it, and if interested, they can also participate and so the cycle continues, gaining more and more visibility for you and your business.

We have seen the type of contests that you can hold to advertise your product; here are a few other things to keep in mind when you *do* conduct an online contest –

- A contest that's conducted randomly is boring and not very successful; set yourself a proper objective. Why are you holding this contest? The purpose may or may not be something you want to make public – that's your choice to make. But you do need to have a definite purpose in mind when you begin.
- When you have a proper goal defined and set up, you can make your posts specific to that goal. You can identify your target customers and then makes posts and edits that will attract their attention; for instance, if we were conducting a contest to promote Body Positivity as our Corporate Social Responsibility Initiative of the Year, we at Awesome Girl Apparel would look for a target audience of the extremely bigger and extremely smaller sized women who face insecurities about their looks. The contest would ask them to post pictures of themselves wearing our clothing and looking fabulous. It could also get personal, explaining if and how our clothing helped them get over their insecurities – without a purpose, this type of contest would be a total bust.
- Again, the purpose of the contest will determine what your prize is. We have already discussed a

bit about what you want to give away and if you're willing to spend extra on this. Why you're conducting the contest and who your target audience is will go a long way in helping you decide what your prize should be. For instance, if we were going with the above mentioned contest at Awesome Girl Apparel, we would probably offer them a day with our designers or a full outfit or look for free.

- Apart from simply announcing that you are conducting a contest, you will need to post regularly about the progress of the contest until the deadline passes. Simply having one post at the beginning will not give you any visibility; too many posts can also get annoying! You should, ideally, post at least once a day, reminding your Followers about it – you should also link to your participants' posts so that everyone can see them. Keep reminding your Followers about what's happening and post regularly about the prize so that more people get to know what's in store for them if they do participate.

- Remember to manage your contest properly by creating a specific hashtag for your contest. If you don't have a hashtag specific to your contest, you will end up floundering like a fish out of water when it comes to managing your entries and your participants – you may even miss out on a few of them! Keep in mind that the hashtag cannot be something that is common or regularly used – you don't want it getting mixed up with some other tag! The tag should create a unique community on Instagram for you and your customers and your participants so that they can all view one another's entries and the purpose of the contest is well served. Also remind your participants to tag your Instagram account when they upload their respective pictures; otherwise, their entry could be missed too.

- Make sure to keep posting the rules and the details of the contest regularly. Also make it necessary that if they want to take part in your contest, they should be one of your Followers – this will bring in more visibility and make it easy for you to keep track of your participants and their individual entries.

- Remember to set a deadline for the contest's validity. Make it very clear that entries after that will not be considered. Now, setting a deadline is a very delicate task – if it's too small a time frame, you won't get the visibility you need. If it's too long, people will get bored and move on or forget all about it. Experience shows that small, regular contests with a time window of one or two weeks are very productive; it generates enough traffic and keeps your customers buzzing with excitement and doesn't require much in the way of investment. If, however, you are giving away a big prize into which you're putting a good amount of money, then a longer deadline is a better option – you will have to make your customers work to receive the prize. This is helpful in building brand loyalty, but it is time consuming and requires some effort on your part.

- Make sure to announce the basis of your winner selection. You want to conduct a transparent, open contest that lets your customers know that you are a clean business. Make the announcement at the very beginning so that there is no conflict or confusion.

- Remember to promote your contest on every social media platform you are active on. Even if it is only an Instagram Contest, specific to the features of that particularly app and social media, promote it to get more visibility for both the contest and the venture. Tweet regularly about it, put it up on your Facebook page, blog about it and on each medium, provide a link to your Instagram account where the contest is actually being held.

- Ask previous contest winners to do a shout-out of your contest on their own accounts; see if they will be interested in marketing for you, in return for something simple like a shout-out of your own.

- Don't let go of your contest once it's begun – you will need to monitor the traffic to your account. That is the purpose of a contest, after all! Use third party tools like Google Analytics, Google Alerts, Twitter Analytics, etc., to see how much visibility your contest has gained you. Also analyze the number of people who have Followed or Unfollowed you during and after the contest – all of these will help you understand and analyze the effectiveness and the impact of the contest and if you should do another contest or not.

- Decide how you are going to give the prize to the winner of your contest – are they geographically close enough to receive it from you in person? Or will you have to ship it to them to their home? If so, then how much is it going to cost? All of these are things you should consider even before you begin your contest so that you can target that specific type of audience.

As you can see, conducting an Instagram Contest is not a small thing; when you make it part of your advertising and business strategy, it requires careful planning and effort to make it worthwhile. But if you do it right, it could get you, your account and your business venture a huge amount of visibility and the prospect of new as well as return customers!

Chapter 7

Establish Clear Guidelines for Your Team: Style, Publishing & Workflow

A consistent voice on social media is key to building your brand, and on a visual platform such as Instagram, the need for a clearly defined aesthetic adds an additional layer of consideration. Even if one person is responsible for managing your brand's Instagram account, establishing guidelines for photo and video composition, filter use and captions will ensure that your Instagram content is part of a unified brand experience for your followers.

Create an Instagram Style Guide

At the point when distributed on Instagram, you have a bigger number of choices to make than which filterl looks best. From visual synthesis to area labeling to utilizing hashtags as a part of your subtitles, arranging your methodology ahead of time permits you to amplify the capability of each Instagram highlight and usefulness. Your style guide should outline your approach to each of the following:

Brand tasteful: Review the current visual representations of your image: your logo, site, illustrations, stock photography and other security. Do you have a built up shading palette? A cool or warm tone in pictures? Your Instagram content, and the altering impacts and channels you pick, ought to mirror the same.

Structure: Not each online networking advertiser is a characteristic picture taker, and shooting for Instagram and for a portable group of onlookers is an educated aptitude. While you can now distribute pictures with a scene or representation introduction to your Instagram bolster, every bit of substance will appear as a square thumbnail in your profile framework. Decide your way to deal with a couple of fundamental components of arrangement keeping in mind the end goal to make a feeling of visual concordance when a client takes a gander at your profile.

- Backgrounds

- White space parity

- Dominant color(s)

SubjectMailChimp is one example of a company with a well-defined brand aesthetic and excellent, consistent photo composition. Its photos, videos and GIFs typically use a simple background with one dominant color, allowing the subject to stand out.

Utilizing Filters, Lux and Creative Tools: Instagram offers a few approaches to alter photographs and recordings. Survey channels and their belongings to choose a modest bunch that fit your image's stylish and guarantee outwardly steady substance.

For photograph altering, you additionally have the choice to apply Lux or use inventive devices. Lux (the sun image) conforms the difference and immersion of your photograph. Apparatuses (the wrench image) permits you to exclusively change brilliance, contrast, warmth, shadows, shading and that's only the tip of the iceberg. For video altering, you can choose a channel, trim substance and pick a particular spread picture that will appear in the News Feed.

Caption: Approaches to composing Instagram inscriptions differ—inscriptions are constrained to 2,200 characters, and subtitles are truncated with an ellipsis after three lines of content. While a few clients discard inscriptions out and out, others approach sharing as a type of microblogging and compose a short story to go with each post. Try to incorporate the most critical piece of your message inside those initial three lines regardless of the fact that you settle on a long-frame inscription. In the event that you require any motivation, we gathered together a lot of innovative Instagram inscription thoughts.

Similarly, as with all parts of style, consistency is vital. Your duplicate rules ought to incorporate whether sentence parts are adequate, on the off chance that you plan to utilize emoji and hashtags (and what number of to use in a given post), and what your approach is around @mentioning different clients.

Hashtags: Hashtags permit Instagrammers to find substance and records to take after, so utilizing them is a decent approach to associate with new devotees and expansion engagement on your posts. Instagram permits up to 30 hashtags per post or remark, so choose whether your image will utilize them, what number of to incorporate into an ordinary post and whether you need to make marked hashtags to adjust to your substance topics. As far as duplicate rules, you can utilize hashtags inside your duplicate—"welcome to #NYC"— or toward the end of a post. Which looks best to you?

Give yourself an opportunity to scan inclining hashtags inside the Instagram application's Search and Explore tab to see what individuals are discussing and discover open doors for your image to join significant discussions. While it's enticing to arrange the greater part of your social posts ahead of time, joining drifting discussions is a decent approach to interface with new gatherings of people and stay consistent with the "insta" a portion of this application—regardless of the possibility that that implies planning time for suddenness now and again.

Add to Photo Map: Location labeling, or geo-labeling, utilizing the Add to Photo Map highlight is another approach to expand engagement and permit new clients to find your substance—posts with an area get 79% higher engagement than posts without. For associations that utilization Instagram amid travel and occasions, at different locales or to advance destinations, geo-labeling is an especially helpful element.

Label People: If you tag other Instagram individuals in a post, it will appear on their profile under the Photos of User segment. You can utilize this usefulness to label people or brands highlighted in your posts, and clients will have the capacity to tap your photograph to view and snap labeled handles.

Social Sharing: Instagram permits you to interface your profile to accounts on Facebook, Twitter, Tumblr, Flickr and Swarm and naturally push your photograph to those systems. You can likewise make IFTTT Instagram formulas to naturally share your photographs in Slack channels, as Facebook collections or on Twitter.

Figure out if you need to cross-post or advance your Instagram content along these lines. On the off chance that cross-posting is a vital piece of your procedure, ensure that anybody dealing with your Instagram account has entry to connected records on the off chance that they have to reauthenticate the association.

Use Landscape: Having spotless, all around edited and expert Instagram photographs is vital to your business. Presently you can utilize Sprout Social's most up to date picture resizing device Landscape. This free device makes it simple to resize, harvest and scale your online networking pictures over all stages.

Recognize Team Members and Roles

Your essential online networking administrator ought to absolutely be a piece of your Instagram promoting, however other colleagues may likewise give profitable commitments. Contingent upon your group and targets, you may separate obligations into substance creation and distributed, group administration, disclosure and investigation, and relegate them to colleagues with various qualities.

For associations that require outbound message endorsement before distributed, set up a reasonable procedure for making and assessing content. One proviso, however: Capturing and sharing minutes progressively is a piece of Instagram's motivation. At the point when posting amid live occasions, ensure your work process will hold up.

Chapter 8

Foster Engagement & Set Guidelines for Community Management

From curating UGC to empowering exchange and building a group, Instagram offers tremendous potential for engagement with devotees. In the event that you stick to distributed and skip engagement, you will pass up a great opportunity for a chance to naturally develop your taking after by interfacing with fans and contacting new crowds.

Optimize Your Bio & Link

With a 150 character restrain, your Instagram bio ought to concentrate on what's most critical about your brand. Your bio is a decent place to teach clients as well: While clients can't tap on hashtags in your bio (on the versatile application) the way they can in subtitles, including a marked hashtag advises clients how to share and find extra substance identified with your image. Putting your best foot forward is a workmanship—on the off chance that you require help composing your Instagram bio, look at these proposals.

Remember that the main live connection you can incorporate on Instagram is in your bio—no one but sponsors can share live connections outside of that space. In the event that one of your destinations is to drive movement back to your site or blog, incorporate the connection in your profile and allude to it in individual posts by utilizing content like "connection in profile" in your inscription.

Following Accounts

What kind of content do you want to keep a pulse on through Instagram? Following influencers in your industry—for example, if you are a clothing retailer, following top fashion bloggers—will help you keep an eye on interesting content and even find inspiration for your own posts. It helps to set basic guidelines around who your brand will and won't follow, taking into consideration a few things: private accounts (the user must approve your request to follow), employee accounts, relevance to your brand and appropriateness of content (nothing that's not safe for work, whatever that means in your industry).

Monitoring Hashtags & Location Tags

Another approach to see who's discussing your brand, regardless of the possibility that they aren't specifying your handle, is to screen marked hashtags. Those hashtags regularly incorporate your image name and any normal incorrect spellings; names of items,

Grant Kennedy

administrations or occasions; and terms identified with your association. For us at Sprout, that implies observing hashtags like #SproutSocial, #TeamSprout, #SproutLove, #SproutChat and #SproutAndAbout. You can't control who will utilize your marked hashtags, so some posts may not be important, but rather observing the discussion will permit you to draw in with adherents who are offering your image to their systems.

In the event that specific areas are huge to your association, for example, your base camp or occasion spaces, frequently checking area labels will likewise help you distinguish clients posting from those territories. At the point when an area is labeled in a post, clients can tap on its name to see all photographs labeled in that place from open records and private records that they are taking after—giving them a review of all substance shared from a specific area. This is what this looks like in Sprout's Smart Inbox:

Search & Explore

Just as you monitor branded hashtags, you can identify popular hashtags in your industry to monitor for engagement and use in your own posts. Join new conversations related to your brand by searching results for the hashtags you use, then examining hashtags used by people sharing similar content.

Analyze Your Results

Following how well your substance performs and your adherent development will permit you to adjust your

Instagram showcasing technique after some time. This permits you to convey a greater amount of the substance that your gathering of people reacts to while enhancing your arrangements for future battles. At the point when taking a gander at your sent messages, dissecting the quantity of remarks and likes got, and also the engagement rate for every post, will demonstrate you how distinctive sorts of substance perform. The engagement rate is a rate of preferences in addition to remarks on the post, separated by devotees of the record at the time the post was sent.

Different measurements, for example, general adherent tally, marked hashtag notice and area labels can give you a thought of the aftereffects of your distributed and engagement endeavors. You can likewise utilize post execution information to evaluate different elements, similar to the best time of day for group of onlookers' engagement. Numerous brands likewise utilize a connection to an Instagram-particular point of arrival or with a particular UTM tag so as to track the quantity of snaps on the connection in their Instagram profile. Sprout Social's Instagram examination can help you track group of onlookers' engagement, distinguish your top posts, find drew in influencers and that's just the beginning.

Managing Comments, @Mentions & Direct Messages

On the off chance that your brand gets an expansive volume of engagement on Instagram, it can be hard to guarantee that no messages become lost despite a general sense of vigilance. Sprout Social's Instagram administration devices can help you screen, draw in and break down your Instagram endeavors. Perceive how it functions for yourself with a totally free, 30-day trial today.

For those messages that emerge as requiring a reaction, your group ought to have an arrangement set up for taking care of certain sorts of messages:

- Support or client administration request

- Negative or deriding remarks

- Career request

- Sales leads

- Spam

Certain messages may require a reroute to another stage; for instance, it might be less demanding to handle bolster issues through live talk, telephone or email. Utilize Bitly or another connection shortening instrument to set up a simple to-recollect short link to pages identified with normal request.

A typical practice on Instagram is to @mention different clients—now and then in an elucidating inscription, now and then in a remark—to give credit or attract their thoughtfulness regarding a specific photograph or video. On the off chance that you get notices that your image is @mentioned, it may be worth spending a minute to recognize an on-brand fan photograph with a Like or a remark.

Instagram Direct permits clients to send a photograph or video specifically to anybody as opposed to distributed it to their food. In the event that a client sends this kind of message to your handle, and you aren't tailing them, the post will be in your solicitations line, where you can acknowledge their solicitation so as to audit and react to their post.

While some brands have effectively utilized Instagram Direct to gather client submitted photographs and remunerate fans, if these messages aren't a piece of your battle or methodology, choose a reaction technique. When all is said in done, keeping associations to open posts works best for generally marks.

Engaging with the Instagram community Engaging with the Instagram community is essential to your success on the platform because your brand is a part of the community. In addition to posting high quality content, you should follow other accounts, comment on and like posts by other accounts, and respond to questions and comments on your posts. We recommend the following:

1. Accounts to follow: Follow your brand's partners, spokespeople, advocates, and influential members of the community whose content may relate to your brand.

2. Commenting & liking: Use hashtags, location and Photos of You to find images posted by other accounts about your brand and engage with them by liking and commenting on their images.

3. Account moderation: Set up policies and procedures for responding to questions and negative comments. Determine how you will handle offensive or inappropriate comments on your content.

4. InstaMeets & InstaWalks: Host an event or walk where you invite the community to meet in person and take Instagram photos and share them with a specific hashtag. Reach out to local Instagram Meetup groups to attend. You can find a list of Instagram Meetup groups at meetup.com/instagram.

5. Weekend Hashtag Project: Participate in Instagram's weekly event when we invite Instagrammers to take images throughout the weekend that fit a specific theme and post them with a specific hashtag. A new project is posted every Friday and a roundup of favorite submissions is posted on Monday morning at blog.instagram.com.

Leveraging user-generated content (UGC) on Instagram The ability to source high quality brand content from the Instagram community is one of the major benefits of the platform. Filters, image effects and the square format help anyone turn a simple snapshot into a work of art. People frequently post images of products or of retail locations and because many also share their posts publicly, you can view them by searching hashtags or locations associated with your brand. People may also use the Add People feature to add your brand to their photos, which enables them to appear in Photos of You in your Instagram profile (pending your approval). We recommend the following guidelines when sourcing usergenerated content on Instagram:

1. Curation: While much of the user-generated content on Instagram is high quality, it will be necessary to review and curate the photos and videos you want to use. Designate a team to handle this internally or consider hiring a third party company to assist with this process.

2. Image attribution: Always attribute an image to the original creator by listing their handle and referencing the source (i.e. "@[username] on Instagram" or "[camera logo] @[username]) on the image itself or in the caption when posting to your Instagram account.

3. Legal considerations: Instagram cannot advise your brand on how to obtain approval for the use of images posted by other accounts on Instagram. We recommend that you discuss with your legal team and establish a process for leveraging content posted by other accounts, whether you are reposting on your own account, a website or in any other marketing materials.

Running a photo or video contest on Instagram Brands often run contests to inspire people to post photos or videos of their brands' products and locations to increase awareness of a product, campaign or initiative. We recommend the following guidelines when running a photo or video contest on Instagram:

1. Use a unique hashtag for submissions. This helps to ensure that the account intended to submit the photo or video for your contest and reduces the amount of moderation needed.

2. Create a photo or video to share on Instagram to introduce the campaign. Share it on Facebook and Twitter to promote the campaign to your followers on those platforms as well.

3. Have a few examples ready to display immediately when you start the campaign. This gives people some direction as to the type of photos or videos they should submit and encourages higher quality and creativity in the submissions.

4. Display a curated set of the photos or videos somewhere (on your website, on a Facebook tab, highlight some in your Instagram feed, etc.). Displaying a gallery shows people that others are participating and will encourage them to do the same.

5. Do not choose the winner based on the number of Instagram likes a photo or video receives. This tends to trigger bad behavior on Instagram (spam comments, auto-following, etc.) rather than encouraging good quality content

Chapter 9

Apps to help you boost Instagram Marketing

Instagram promoting... it began much sooner than Instagram acquainted advertisements with the stage. Be that as it may, it's just been done well by a chosen few brands. Furthermore, to call it as is it, a large portion of those effective brands are as of now outwardly determined way of life brands: Anthropologie, Target, W Hotels, Coca-Cola, and so on. It's just from time to time that you go over an ought to be-exhausting brand with a totally dynamic Instagram nearness. (I'm considering General Electric as a fantastic illustration.)

Along these lines, suppose you're one of those outwardly exhausting brands. You know... your organization offers administrations or propane. As such, you don't simply normally have a mess going for you on the stylish front. On the off chance that that portrays your organization, then these five applications for showcasing on Instagram ought to be in your download line promptly!

Iconosquare

Iconosquare is an unmistakable and valuable stage to help you to deal with your Instagram account, making it less demanding to associate with different clients and screen your record. Sign in with your Instagram login subtle elements and explore between the tabs on the top. Not just will it be less demanding to see your posts however the stage additionally gives details. The key information to take after are "Adoration rate", "Talk rate" and "Spread rate", as they are a decent sign of how you're performing.

Hit the "Improvement" catch to get bits of knowledge into what your supporters are keen on, what the best times to post are, filter affects and everything in the middle.

On the off chance that you tap on "Advance", you will see a rundown of choices to advance your Instagram profile. You can interface it to your Facebook page, make a spread picture utilizing their generator, and so forth.

Hyperlapse

Hyperlapse captures high-quality time lapse videos. It features built-in stabilisation technology to create moving time lapses, giving a cinematic look to the video.

To record a video, just launch the app - you don't need an account. Tap once to begin the recording, tap once again to stop. Then speed up or slow your video down by choosing a playback speed between 1x and 12x and tap the green check mark to save it. Once this is done, share it on Instagram. This app will help you to create professional looking content for your brand on Instagram, so be creative.

Offerpop

Offerpop is a marketing tool that lets you create and manage campaigns across just about every social media channel you can imagine (with the exception of Google+). You can track Instagram hashtags, track performance, store user-generated content in an easy-to-access library, knock out auto replies for your campaign, and publish photos across different channels. It's also worth noting that brands like MTV, Pepsi, TOMS, Gap, and American Express use Offerpop.

Slidagram

Jumping over into the world of *free* apps for marketing on Instagram, Slidagram allows you to create slideshows on your website that are based around popular hashtags or location. For marketers with a huge library of Instagram images, this is a great way to easily extract the photos that pertain to your upcoming event or campaign. Plus, it's hard to argue with free.

VSCO Cam

Short for Visual Supply Co., VSCO allows iPhone and Android users to create striking images, using unique filters that are almost in direct opposition to Instagram's many heavy-on-the-saturation filters. Maybe your brand will find its aesthetic in this app. To learn more about the quickly growing company, I highly recommend this in-depth read from Fast Co.

Afterlight

Afterlight is kind of like VSCO Cam's next-door neighbor. Combine the editing features of the two apps and you'll find yourself headed off in the general direction of aesthetic bliss. Especially valuable: Afterlight's 75 unique, simplistic, and adjustable frames.

Followgram

Followgram is the ultimate in Instagram analytics. This is a must-have app for marketing on Instagram. Followgram provides tons of free stats about who liked and commented on your content. Plus, the service offers special tools that allow businesses to run contests and get accurate analytics. Followgram is used by Ford, National Geographic, Nike, Armani, and others.

Chapter 10

The 5/3/1 rule

Racking up followers is no easy task. To help you doing so, you need to reach out to people. A technique that you can use to gain followers is the 5/3/1 rule:

1. Start by finding a user in your niche.

2. Like five of their photos.

3. Comment on three of their photos.

Result: Earn a new follower!

If you use hashtags correctly, users will find your posts. Exchange likes for likes, and be sure to leave a reply to commenters.

Instagram does not have a "share" feature, but apps

like Report for Instagram enables you to curate users' photos.

Instagram Etiquette

Being a "social" media site, there are of course some etiquettes to consider on Instagram.

Following People

If you start following someone on Instagram, don't just follow them and like their latest photo. Go through their gallery and like other posts. Comment on at least one of their photos. You can even mention that this post inspired you to follow them. Reaching out to them like this will establish a strong connection from the start and they will often follow you back as a result. Keep the relationship growing by continuing to like and comment on additional posts in the future.

It is also not acceptable to follow a bunch of other people in order to get them to follow you and then turn around and unfollow them. This is not how you grow an audience! Follow people who actually interest you not just to build likes.

Sharing Photos

As the popularity of Instagram increases, I'm finding more and more accounts taking the "easy" road to sharing images. Instagram is all about sharing your own images! The purpose of Instagram is to share images instantly - right when you take them on your phone. Hence the "insta" in the name!

Grant Kennedy

I understand that we might not always share them instantly. In fact, I often recommend waiting and spacing out your posts. And I know many photographers who use their DSLR cameras to take photos then format them for Instagram upload after the fact. I am totally ok with both of these practices.

However, I am not ok with Instagramers using other people's images, or images they found on Google.

Instagram images should be a reflection of you! Using other people's images is not a reflection of you.

Many of us get in the habit of sharing inspirational quotes or seasonal images that we find in searches. It's extremely common on Facebook and some of the other sites. I get it, it's easier than creating our own. And sometimes, we find something that just resonates really well and we want to share it.

But Instagram isn't really about this behavior. If you want to share an inspirational quote, put it on your own image! Take a few extra minutes to find a good photo from your gallery, add text for the quote, and then upload the formatted image to your Instagram account. There are a variety of apps you can download that add text to images.

You should also know that a lot of Instagramers are extremely vigilant with their copyrights. They will call you out if they find you using one of their images without attribution or credit. Trust me when I tell you this is not the impression you want to make on Instagram!

You should also consider the frequency with which you share posts on Instagram. No one wants to go through their feed and see 25 photos from you, one right after the other. Space out your images to ensure that other users' photos fill the space between yours. Not only will this keep your photos active throughout the day(s), it will also increase your engagement on each photo. If you share 10 photos in a row, by the second or third one, people stop liking and commenting. However, if you space them out and keep them fresh, you're more likely to continue getting likes on comments in higher proportions on each post.

Also, share only the best photos with your audience. Just because you have 3 different angles of that birthday cake doesn't mean you have to share them all. Choose the best one and share that one.

Tips and tricks

The best days to post are Sunday, Monday and Thursday

-The average number of hashtags for the optimum engagement is 11

-Use a mix of brand specific hashtags, industry specific hashtags and trending hashtags for best results

-Share your photos across every social platform

-Follow influencers, engagers and other brands to extend your reach

-Make sure you have at least 5 images in your Instagram stream before you begin following others

-Add questions in your caption that encourage comments

-Reply to any comments. Be sure to @mention the commenter in your photo

-Like and comment popular photos. @mention both the creator AND the commenters

-Use Instagram Direct to contact top followers

INSTAGRAM + BEYOND

Connecting your Instagram account with your accounts on other platforms

The ability to cross-post to other online platforms is a unique feature of Instagram's platform. We recommend linking your Instagram account to your accounts on other platforms. Sharing content from Instagram to other platforms lets people on those platforms know about your account. However, you should develop a strategy for determining how often and which posts to share to other platforms according to your objectives. For detailed steps on how to connect your Instagram account to other platforms, refer to the guide Connecting and sharing from Instagram to Facebook & Twitter.

Embedding photos and videos on other websites Instagram provides the ability to embed public photos and videos from the platform on your site. We recommend using this feature whenever the images will enhance content on your website or as a way of letting visitors to your website know about your presence on Instagram. The embed feature automatically gives credit to the original creator of the content, showing the image or video with the username. For more on embed codes, visit the Instagram help center at help.instagram.com.

Leveraging the Instagram

API The Instagram API can be used for a wide range of creative executions, many of which haven't even been imagined yet. The API allows you to pull photos and videos from Instagram and display them on another website, digital interface, or even physical objects, in the real world. Photos and videos can be accessed by user, hashtag, time, or place. For more about the API, visit the the Instagram developer help center at instagram.com/developer.

Here are some of the best uses we've seen of the Instagram API:

• Pull selected photos and videos with a specific hashtag into your site to show real people using your products or attending your events.

• Ask people to submit photos or videos for contests or campaigns and create a place online to display them.

Grant Kennedy

• Create a digital map of photos and videos posted by your account and/or other accounts posting about your brand.

• Create a large screen or multiscreen display that pulls photos and videos from Instagram in real-time during live events.

• Print photos to use in displays or to give to attendees at events.

Working with API partners Many third-party developers offer tools and services to manage and monitor your Instagram account beyond what is currently available from Instagram. These service range from account management and insights to content integration.

Using the Instagram brand and trademark

It's recommended using the Instagram brand and trademark in your communications to partners and customers to spread the word about your Instagram account.

Find guidelines to ensure legal compliance when using our brand and trademark at help.instagram.com.

Using Instagram screenshots

Should you want to include Instagram screenshots in your marketing materials, please follow these guidelines to ensure legal compliance:

• Screenshots must be unaltered, meaning they cannot be annotated or modified in any way from their appearance on Instagram.

• Screenshots with personally identifiable information (including photos, names, etc. of actual users) require written consent from the individual(s) before they can be published.

• Screenshots of any user photos will need written consent from its creator before use.

• Please include: "All Instagram™ logos, marks and symbols are the property of Instagram, Inc. Copyright 2010-2013. All Rights Reserved."

Chapter 11

Some Quick Marketing Tips for Instagrammers

So far, we have discussed all that you can do by yourself on Instagram to promote your product and establish your brand online. all of this was done without having to pay more than a few dollars; in this chapter, we will take a look at some of the paid options to promote yourself better.

When they saw how well Instagram could be used for business, the owners decided to set up a service for advertising via Instagram. Instagram Advertising is something you will have to pay for, though the cost is still very less compared to what you might pay for an ad film otherwise. Here is how you can use it –

- Sync your Instagram account to your Facebook account. The advertisement you will be making will be created with the Facebook Power Editor tool, so log into your Facebook account and then under your settings option, click on the Instagram Ads tab. Here, you will be asked to add an account if you have not already synced them – enter your username and password for your Instagram account and then you're ready to go!

- Now you need to pick the type of advertisement you want to make. The options you are given are divided according to the purpose of your ad campaign; they could be video views or clicks for your website. Here are your options –
  - You could pick an image advertisement which will simply have a picture, along with a smaller button that the users click on. The button will link to the website they can get more details at or make the purchase at.
  - The second type of ad is the video ad, where you will make a video that has the same linked button.
  - The third type is the Carousel advertisement where you can offer your users a series of images they can skim through before making the purchase.
- Choose the type of ad that suits your requirements for you business; the next step will be to set the target audience list. You will have to key in details about who they are, where they are located, their age, gender, occupation, etc. You will be given a lot of detailed options to choose your target audience to make your ad

> very specific to their needs, so make sure you give all the details properly.
- Now you have to create your advertisement's content! Make it creative, make it eye catchy and colorful; if it is a video, keep it simple, keep it easy to understand and maybe even have a trademark jingle to identify your product by.

Now your ad is ready to go! Post it on Facebook where people will further share it and then link it to your Instagram, where you can tag your favorite customers and use trending hashtags to promote your brand!

Another idea is to make use of sponsored advertisements on Instagram. Sponsored ads are those that you pay for. These sponsored accounts exist only to give shout-outs to business entities that are willing to pay them for it. They have a large number of followers and can reach out to a lot more people than you yourself could. Their sole purpose is to further direct traffic to business accounts, so if you are willing to pay them a fee, then they will help you build more traffic to your account as well.

Connect with other businessmen and business owners on Instagram and other social media accounts. Remember, these are social platforms first and foremost – if you want to gain more visibility, then you will have to interact with people, follow them, like their posts and give them shout-outs in return for your own. When it comes to business ventures online, you could partner up with someone else and help them promote their product while they return the favor. For instance, at Awesome Girl Apparel, if we get a fashion designer to endorse one of our designs, we would have to do something for them in return. So we could promote their next fashion show for them on our account, while posting pictures of their endorsement continuously to get the quid pro quo out of it.

Speak to popular Instagrammers. There are some who have a lot of followers and a very powerful presence within the portal; get them to endorse you if possible, maybe in return for something. You could pay them or you could offer them a free sample of your product – their influence as Instagram users and bloggers will go a long way in convincing prospective customers that they should try out your product.

When you post on Instagram, don't limit yourself to pictures of your campaigns or your product alone. In fact, if you can your entire team of workers involved, it will go a long way towards convincing your customers how good and worthy your brand is! Keep your customers updated about your employees' lives as well as your own – this way, a personal relationship between you is formed that goes a long way towards building brand loyalty, since the customers feel like they know you personally.

Post pictures of your team building efforts; post pictures of the production and creation process. For instance, at Awesome Girl Apparel, we would probably post pictures of our basic designs (protecting proprietary information, obviously), pictures of our workers getting the clothes together, pictures of our workers trying on our clothes and then our models promoting them. But apart from that, we would also post pictures of team vacations, employee galas and get-togethers, meetings with regular customers, etc., to show the followers that we hope to build a *community* that we can support and expect support from in return. It's this relationship building that should be your endgame – that will get you the most loyal customer base possible!

Chapter 12

20 mistakes you can't afford to make in Instagram Marketing

20 Mistakes:

1.There's No Link in Bio to Drive Traffic

• The less demanding you make it for supporters to find your store, the more traffic you will get.

2. There's No Description in Your Bio

• What defines an epic bio depiction? - portrays the "why" or mission/theory behind your items - succinctly depicts what you do/offer.

3. Crappy Resolution Photos Are Scaring Away Potential Customers

• Make the determination of your photograph 2x the suggested measure so when it gets compacted regardless it looks fab.

• pictures ought to be 1280px by 1280px

4. Your Photos Aren't Sized Right

• So ensure - your pictures are 1280px by 1280px - remember the square introduction when taking your photographs. - If you have non-square pictures that you would prefer not to harvest, you can utilize an application like Instasize (for iPhone) , Nocrop (for Android) to legitimately estimate them for Instagram.

5. Unpleasant Lighting is Downgrading Your Image

• Good lighting is the distinction between resembling a tween taking crappy wireless pics and a master shooting for Anthropologie indexes

• Try this $12 studio quality lighting set up:

6. Where the Ever-Important Lifestyle Photos At?!

• it's insufficient to simply have item photographs. Those are useful for concentrating on item subtle elements, however you ALSO require way of life photographs demonstrating your items being used. Why? Since they give setting and your clients will have the capacity to envision themselves utilizing your items!

7. All Your Posts Look Exactly the Same
• Photograph your items with various foundations.
• Add way of life photographs.
• Incorporate other substance that is applicable to your gathering of people (not as a matter of course item photographs).

8. You're Posting At All the Wrong Times
• Don't post when your kin aren't on Instagram! Post when they ARE.
• If you arent beyond any doubt when your group of onlookers if most dynamic, take a stab at utilizing Iconosquare to see your measurements.

9. You're Posting Inconsistently
• So being steady it key. 1-2 posts for each day is strong, the length of you do it consistently!

10. Not exactly Engaging Copy Isn't Strengthening Your Images
• Your duplicate ought to supplement your picture and back it up with any significant and critical information you ca exclude in your pictures.
• If more engagement is what you're searching for, then take a stab at asking an immediate inquiry.

11. There Ain't No Call-to-Action in Your Copy

• IF you need your supporters to accomplish something, you're going to need to flat out let them know. (i.e. On the off chance that you need them to visit your store, say "Tap the connection in bio to shop this look!".)

12. Your Copy is Way Too Long and Distracting

• Your post depictions are not intended to be long! Keep them as short as you can while as yet offering the key data or you'll lose individuals simply like that. General guideline: Don't make your peeps need to look to peruse!

13. Being Too Salesy and Killing the Fun, Social (and Beneficial) Part of Instagram Marketing

• The main turn off for potential adherents is specifically requesting that they tail you or purchase when you're expressly captivating.

14. Not Using Hashtags for Exposure Instagram posts with 11+ hashtags get the most collaboration.

• Do some hashtag research on Iconosquare to find the applicable ones to your group of onlookers and post 4-5 as a remark on your posts. Following 30 minutes, erase that remark and compose another one with 4-5 new important hashtags. Along these lines you get the presentation of loads of hashtags without looking spammy.

15. Insignificant Content Turns Away the Right Audience

• If you post the substance your intended interest group needs to see, those are the sorts of individuals that will tail you – not simply llama significant others. What's more, when you assemble a solid, pertinent gathering of people, that is the point at which you can change over them to Customers.

16. Anticipating that Followers should Just Come to You

• Use the hashtags your adherents use since then your pictures will be arranged with the photographs they are now posting and taking a gander at. Use Iconosquare to do some hashtag research.

• Search those hashtags and remark on the general population that utilization them (theo retically these ought to be your intended interest group). Compose decent remarks, "Diggin' this photograph!", and so forth. They will welcome the way that you discovered them and took an ideal opportunity to draw in with them... and that thankfulness frequently means adherents.

17. You're Running Contests When You Don't Have an Audience Yet

• Focus on developing your Instagram gathering of people before you have a go at running a challenge, and after that hit it hard when you have the right crowd!

18. Your Incohesive Look Makes Your Brand Unrecognizable

• You require your image to be conspicuous, and be considered as expert as the greater brands – so you require a firm search for your Instagram.

19. Why Are There Selfies?! It's NOT About You

• in the event that you are an item based business, selfies truly have no spot here.

20. Taking after Way More People Than Follow You

• If you are taking after 5,000 individuals and just have 500 adherents, that makes you look terrible. You need to look believable, and your devotees let others realize that they go to YOU for your substance. Not the a different way.

It is safe to say that you are committing any of these errors? On the off chance that you are, FIX THEM RIGHT NOW! Your Business will bless your heart.

Conclusion

Thank you again for downloading this book!

Social media is the singular, most powerful tool in reaching out to a large number of people. If you do not make it a part of your marketing strategy, especially if you are a small business venture, then you will find yourself facing a huge loss in the Digital World of Today, where everything runs on hashtagging and commenting on a post.

Instagram is a mobile photography platform where you must make use of pictures to tell your customers who you are, what you sell and why they should try your product. A picture is indeed worth a thousand words – keep them fun, interactive and cool so that your customer base grows day by day. Use all your social media accounts, from Facebook to Tumblr to promote your brand – interact with your customers and analyze the trending tags in the market to examine what they want and plan your business model accordingly.

Remember that the key to making your Instagram activity effective is engagement and reach. As important as gaining followers is, it's as important to maintain them. This can be achieved with quality content and interaction with your followers.

Make sure that you are consistent in updating your profile and keep things interesting, but relevant to your brand.

The Iconosquare and Hyperlapse apps can be used to help monitor your Instagram usage and produce creative content, so they're well worth giving a go!

If you keep all these fundamental tips in mind, you're on track for success on Instagram. Good luck!

Thank you once again for choosing this book and I hope that you found it useful!

Grant Kennedy

Grant Kennedy

www.ingramcontent.com/pod-product-compliance
Lightning Source LLC
Chambersburg PA
CBHW060349190526
45169CB00002B/534

* 9 781533 135346 *